ROAMING FREE

ROAMING FREE

WILD HORSES OF THE AMERICAN WEST

Text and photography
by Skylar Hansen

NORTHLAND PRESS  FLAGSTAFF, ARIZONA

To Angie, for her patience while I was away tracking the horses

Until several years ago I thought I knew horses well. They were the calm animals of lush, white-fenced pastures. They were the pampered brood mares at a racehorse ranch where I had worked as a teenager, the gentle mounts ridden by youngsters along woodland paths near my home.

Yes, I knew horses well enough—they were creatures of man.

But all of my certainty evaporated on a late-spring afternoon in 1980. During the course of a wildlife photography trip, I detoured to visit the Pryor Mountain Wild Horse Refuge in southern Montana. And there I saw my first wild horse, a stallion as untamed and defiant as the rugged, barren hillsides where he made his home.

He emerged without warning from a steep ravine, his hoofs loud on the rocky soil. He halted before me, his head held high, his hide rough with scars. We stared at each other for long motionless seconds.

My thoughts clouded as I realized that this was not the familiar animal I had known before. This was a stranger so discordant with my ideals of the horse that I felt as if I had just discovered a new species of wildlife. His nostrils flared as he tested my scent in the air and then took a stride forward. His heavy muscles rippled. I felt the touch of irrational fear, and may have taken a hesitant step backward. Then, with a snort of alarm and a toss of his mane, he whirled and disappeared back down the ravine. I hurried to the edge and watched him drive his band of mares and foals deep into the distant hills.

Since that day I have been fascinated with wild horses, devoting my entire photographic effort to documenting their untamed existence.

1

For months at a time, I've tracked them across sagebrush deserts of the American West. Weeks would pass when I did not see another person, so remote are many of the places these horses inhabit. Yet, every moment has been rewarding. Each new sighting is always as thrilling as the last, many times worth the minor hardships of the back country.

The following pages will transport you into the world of wild horses as they exist today, living treasures of our past. Theirs is a story better told through pictures than words, for words alone can never fully capture the fury of battling stallions, the spectacle of a great herd in stampede, or the evanescent sight of horses silhouetted against a flaming sunset.

I hope that sometime in the future you, too, shall have the opportunity to witness the beauty and spirit of these magnificent creatures.

Wild horses as we know them today have lived in America for only the past four centuries, but the ancestral forms of the horse have a history on this continent dating back through the ages. North America is in fact the primordial cradle of the horse. The earliest ancestor, tiny *Eohippus* (dawn horse), appeared here some fifty million years ago. Scarcely a foot in height, and four-toed, its fossil remains have been unearthed in Texas and Wyoming. Later forms continued to evolve and roam over what are now the western and plains states, steadily growing in size and gradually devel-

oping modern equine characteristics. By the beginning of the Ice Age, about one million years ago, what is now recognized as a modern horse came into existence. It flourished between our shores amid camels and mastodons, and was stalked by saber-toothed cats and giant wolves. *Equus* spread around the world: grazing to the tip of South America and crossing the Bering Strait land bridge to enter Asia. Then, about eight thousand years ago, the horse vanished from the Western Hemisphere, part of the great and still-unexplained wave of extinction that forever swept away so many of the ancient mammals.

But in the Old World, the horse lived on; many centuries later, when hoofs once again touched the Americas upon the arrival of European explorers, it was a moment of reintroduction to the ancestral homeland.

The return to North America began in the sixteenth century with the landing of the Spaniards in Mexico and their subsequent northward explorations. The Spanish horses were essentially an Arabian-Barb mixture from North Africa: compact, tough, and swift in comparison to the heavy mounts bred to carry armored knights of medieval Europe. The ability of these Spanish horses to survive months of inactivity, little water, and poor feed on ocean voyages testifies to their remarkable stamina.

Problems with transatlantic shipping led the Spanish to establish breeding ranches in the West

Indies. It was largely stock bred on these ranches that carried the Conquistadors along the Mississippi River and into the Southwest during the early 1540s in search of fabled golden cities.

The unfortunate Indian tribes they encountered, like the Aztecs of Mexico, were no match for the well-armed soldiers, and were enslaved or slaughtered by the frustrated Spaniards as the latter searched in vain for treasure. After recovering from their initial terror of the mounted invaders, the Indians retaliated. Stealing the Conquistadors' horses became a favorite surreptitious tactic—and thus the return of the wild horse began.

The symbol and property of their enemy, the first horses stolen were often killed in vengeance; the Indians failed to realize the great value of these new animals. It is likely that others escaped or were released. Though an exact place and date cannot be pinpointed for the event, it is felt that in this manner a few of the horses from the exploration parties of De Soto and Coronado were the first to run wild again on the continent in eight thousand years. Many more would have to escape the white man and be dispersed before the horse would become a significant feature on the landscape.

During the 1600s, scattered Spanish colonies were established in the Southwest, and horse stealing became a way of life for the Apache and other tribes. Horseflesh, easier to obtain than buffalo meat, became a favored food. Soon, however, the Indians gained an understanding of

4

horsemanship and developed their unequaled riding skill; horses became prized possessions. Settlements and ranches were raided unceasingly as every brave sought to own as many mounts as he could acquire. Soon, the horse would have the opportunity to be firmly established as a wild animal.

It seems ironic, considering the high value they placed on the horse, that most Indians were at best casual in caring for their stock. But if horse stealing was an art, a test of a warrior's stealth and skill, then horse tending was a chore often left undone. Those horses that wandered into the prairie vastness could easily be replaced on the next raid. It was this combination of a relaxed attitude toward their animals and a brazen confidence among the southern tribes that allowed thousands of the Arabian-Barbs to follow their natural instincts and return to an untamed existence.

The profound effect horses had upon Indian life led indirectly to further dispersion into the wild. Through the late seventeenth and early eighteenth centuries, what has been termed the "Indian horse culture" was spreading outward from the source of the horses. The horse was traded northward from tribe to tribe, and the Plains Indians became transformed from primitive foot-hunters struggling for survival to masters of the buffalo ranges. They were able to hunt the big shaggy beasts with ease from horseback. No longer were all of their energies taken up in the hunt for meat. Now there was

time for sport, for the further development of culture, for expansion of tribal territory, and for war. As the horse brought about this evolution, it became a status symbol in an Indian's life, a form of money in matters of trade, a fixture of myth and legend. The horse had become so ingrained in their culture that the belief arose, particularly among northern tribes who had no contact with the Spaniards, that horses had always lived on their lands and had been a part of their world.

Thus wild horses spread and repopulated the

5

western grasslands from Canada to Mexico. Once an Indian mount escaped it was almost certain that its offspring would remain roaming free. For rather than capture wild horses and tame them, many Indians preferred to trade or raid to acquire new mounts already broken.

By the middle 1800s, immense numbers of wild horses roamed the land. Their peak population has been estimated at from two to as high as seven million. Texas alone is thought to have had one million. Thousands were reported to have been sighted from a single vantage point. In the span of less than three centuries, wild horses had returned in profusion.

Highly gregarious, horses maintain a social order based on competition among stallions to gather harems of mares. Typically, a single dominant male collects from one to eight females, and together with the young offspring he has sired, forms a family band, the center of wild horse society.

The stallion is master of his band, loyal to his mares, and demands their obedience. With gestures he commands them, guiding their movements, regularly herding them into tighter formation as he guards them from rival males. Any female slow in responding to his orders is quickly punished with a sharp nip on the flank. When danger threatens he sounds the alarm, signals the retreat and follows at the rear in order to place himself between the danger and his mares, and to hurry any stragglers. Through frequent confrontation and occasional battles with other stallions, he defends them as his own.

Unlike many large, hoofed mammals—such as deer, where bucks are motivated by the herding instinct only during the rutting season—band stallions maintain their harems throughout the year. This behavior suggests that a stallion's need to possess mates is stronger than the actual mating urge (wild mares come into season only once a year, in the spring, one month after giving birth). Also, though rivalry between competing males is greatest during the breeding period, it continues year-round.

A dominant stallion allows each of his mares to leave the band on only one occasion, for a brief period during her foaling, then she returns as soon as her foal is steady on its legs. I have seen a mare in the company of her band while still trailing the birth sac.

Within each family band, one mare will be most dominant. She assumes a function vital to the band's survival. During an escape from danger, it is her task to choose a safe course and lead the retreat, while the stallion is preoccupied as a rear guard. Lesser mares run behind the lead mare in order of their rank. Any lower-ranking female attempting to pass one of higher rank will be directed back to her appropriate place.

A high-ranking mare enjoys certain privileges. For example, she and her offspring are the first to drink at a small water hole. She will display threat gestures—flattening her ears and threatening a bite—at a lower-ranking female

that crowds near. At times, older mares appear jealous of young harem sisters; they try to remain closest to the band stallion and motion for others to keep a respectful distance. If two mares are in heat at the same time, the more dominant one will show her anger and resentment if the other is mounted first.

Just how each female establishes her rank within a band is not always clear. Certainly age and experience play a major role. But it is also believed that the daughter of a high-status mare, having learned from the behavior of her mother, will later tend to establish herself as dominant over others of her age. Fighting is extremely rare among mares and plays no known part in deciding rank. When a new female is added to the band, she will exchange breaths with the established mares as a part of the process of determining her rank within the harem. The lead mare may almost snort into her nostrils as the new member replies with more gentle exhalations and low nickering.

A family band is the most stable social structure among wild horses, but each band is subject to both regular (concerning offspring) and irregular, sometimes drastic, changes in its makeup.

Contrary to a once-popularly held belief, horses in the wild do not seem to inbreed. At the time of a female offspring's first estrus, the stallion drives her from the band if she hasn't already departed of her own accord; she will be promptly accepted into the band of a neighboring male. Sired colts depart when older, as two- or three-year-olds. However, their ouster is less unceremonious. As a preliminary to their departure, the stallion engages his sons in a ritual of mock battle display, imparting early training in the fighting skills the colts will later need in order to establish harems of their own.

On an irregular and fortuitous basis, the number of mares in a band may be altered at any time depending on the strength, skill, age, and aggressiveness of the stallion. It is an inevitable fact that every dominant male must one day be deposed, and the species carried on, by a younger challenger. Old, battle-weary stallions most often pass the last years of their lives in solitude.

A second major aspect of wild horse society exists in response to the uneven distribution of mares. For every powerful stud able to win and hold a bevy of mares, there are large numbers of younger stallions left mareless. These juveniles run together for companionship in young stallion bands. Like adolescent humans, they are between two worlds: outgrowing colthood, but still too immature and inexperienced to compete successfully against elder males. No equivalent social group exists for fillies, who pass directly from their dam's side into their adult role.

Young stallions, boundless of energy, inquisitive but unsure, restless roamers, establish close bonds in order to fulfill their need for sociality. For the most part, theirs is a carefree existence; unfettered by the responsibility of a harem, their antics are a delight to watch. A quick bout of play, a gallop to the water hole, and then a pause to graze, or a session of mutual grooming of unkempt manes, or perhaps a doze side by side in the afternoon heat. But their ofttimes insecurity is apparent, too, whenever they huddle together like foals at their mothers' flanks, in the face of something new and strange (like a crouching photographer and the click of the camera shutter).

Within a young stallion band an especially close bond often exists between one or more

pairs of the youngsters, like best friends within a gang of boyhood companions. These duos are easily picked out from the rest as they travel shoulder to shoulder, groom and play together, and mostly ignore their other companions except to remain with the group.

Typically, each band consists of two to six members. Since a stallion will usually be five to eight years old before taking a mate, up to a quarter of his life is spent in these bachelor groups. During these years of maturing, as herding instincts surface, a shift from the

9

frolicsome to more serious and aggressive behavior occurs. Fighting skills are honed through mock battles with companions. The oldest or most dominant bachelor stallion will establish himself as the band's leader. He may herd the others and defend them against the friendly approaches of other bachelors as if they were his mares. Such defenses can turn as violent as a battle over a female. He will begin to make bolder challenges to the dominant stallions he encounters, for the hunger to possess a band of mares, the purpose of his existence, becomes an obsession.

Heightened instincts can lead to surprising brashness and uncharacteristic behavior. While other horses flee the approach of a car, a mare-hungry young stallion may draw near, prancing and snorting before it, seemingly willing to test foes of any possible description.

Perhaps as many first mares are taken by a pair of bachelors, fighting in a concerted effort, as those taken by loners. Although an effective means of outflanking a seasoned foe, the result is a small harem lopsidedly in favor of males—a tenuous situation at best. For a time, relative peace reigns between the longtime friends; if the mare is in heat, both will cover her. Sooner or later, however, a fracture is certain, though the companions may not actually separate for some time.

Despite the usual rule of a harem led by a single stallion, it is not uncommon to find multiple-stud harems. I suspect that a large proportion of these relationships occur when, for whatever reason, bachelor companions remain bonded after mares are taken. The balance may stem from a challenger forcing his way into an existing family band, with neither of the two males able to rout the other. In either event, one will emerge as dominant (claimer of breeding rights) and the other subdominant. Each will play differing roles in the security of the band.

The overpowering concern of the dominant stud will be the continued presence of the second. As the band grazes, travels, or rests, he will persist in positioning himself between the mares and the subdominant male. During the course of a day, the two will encounter and briefly display threats a score of times: confronting, half-rearing, and backing apart. There may be periods when open hostilities break out, intermittent battle occurring for several consecutive days. But surprisingly, once the second-ranking male acquiesces and returns to his social position (or perhaps the fighting ends with a switching of roles), then the dominant animal often seems content to allow the defeated to remain with the band.

Having preoccupied the dominant stallion with the rival in his midst, the subdominant horse is the first to deal with outside threats. It is he who rushes off to confront approaching male intruders before they can reach the mares, while the dominant stallion remains with the band as a second line of defense. Only if the intruders are

not discouraged by the subdominant stallion does the dominant horse stride imperiously into the dispute. Thus, though it must be an unfulfilling role for the second stallion, and a burden for the first, such a duo does present a stronger barrier against mare-questing foes.

A third form of grouping, a loose-knit and often shortlived social organization, occurs when a number of bands cluster into a large concentration of horses, forming a herd. A century ago herds were commonly reported to total in the hundreds, even thousands of individuals. But the more modest numbers of wild horses today tend toward collections of no more than sixty to one hundred fifty animals, from several to as many as twenty bands.

Such herds are not to be found on all wild horse ranges. Where horses are few, as in the Pryor Mountain Wild Horse Refuge, bands are inclined to remain separate and to establish specific territories, in which the strongest dominant stallions hold the largest and best pastures. For a herd to form, horses must number in the hundreds on a given range. Not all band stallions will choose to risk their harems near potential adversaries; half or more of the proximate bands will shun the herd and roam independently.

The irregular, impermanent nature of a herd must be remembered if one is to obtain a clear impression of its occurrence in wild horse society. A government wildlife specialist once assured me that herding is predominant only in wintertime, yet I have observed at least as many in mid-summer. Lasting for weeks or months, a herd may disband at any time, to reform later for reasons understood only by the horses themselves.

At first glance, a herd appears as a sheer mass and mingling of horses, formless and without order; a definite structure soon reveals itself. Family bands are by far the more numerous in a herd's composition, outnumbering young stallion bands four or five to one.

Though an entire home range is shared by all of the horses in the herd, a form of territoriality is still maintained by the band stallions. No commingling of mares from other harems is tolerated for long. Each stallion works to keep his band bunched and separate from those surrounding it. As the herd moves while grazing, the individual territory of each band also moves, much like flying birds that keep a wingspan apart to avoid colliding with one another. Should one band approach too near its neighbor, both band stallions will meet on the narrowed buffer space between, begin to confront and posture, and then direct their harems to veer apart. Males voicing their threats and bands shifting course are a regular part of the herd's atmosphere.

A pecking order exists among the many dominant males present within a herd. This is most apparent when the herd is in flight: only one stallion will hold the high-status position of last rearguard. There is also a ranking among individual harems, as is the case among the horses within each band; it lends a margin of stability to the whole herd. So long as no dominant stallion attempts to usurp the relative position of another, a truce seems to exist between the leading males.

Young stallions usually keep on the outskirts, thereby avoiding confrontations with powerful elders. Those that venture inside the herd must be very careful to maintain their distance as they move between the harems; to do otherwise would risk inviting a prancing aggressor. Sometimes in the course of play, an inadvertent infringement on territory will draw a quick response. Regular bachelor members of the herd, who show submission in their own gestures, are invariably allowed a graceful retreat with only a perfunctory warning when confronted. But the same cannot be said for those who react with aggressive posturing of their own. For a stranger who doesn't yet know his place, the welcome is certain to be swift and unpleasant.

Such a mareless male may be approached, threatened, and driven away by a succession of band stallions as he appears to ricochet through the herd from one confrontation to the next. Finally he will be chased to the outskirts and beyond by a particularly enraged dominant male. The chase can last for half a mile, the pursuer nipping at the youngster's rump and parrying upward with his chin, as the pursued lashes back with both hind hoofs every few strides.

I have witnessed herding in Wyoming on Chain Lakes Flat, below Atlantic Rim, and in the Adobe Town district, as well as in the Owyhee Desert and Stone Cabin Valley in Nevada, all areas containing large numbers of wild horses. A wide valley or a plain seems a favored environment for this massing behavior. A good pasture or proximity of a limited water supply probably causes the initial assembly of the horses, followed by loose bonding as an outgrowth of their social nature. Herds are most impressive in stampede. It is also an awesome

sight to see a herd traveling in a long string, band following band, horses moving single file along a narrow trail worn fetlock-deep in the dust, on its way to a water hole or shifting to a new grazing area.

For those bands that normally remain independent of any gathering, a temporary form of herding can be observed when the horses are faced with danger. In a retreat, they may join with the main herd if it is nearby, or combine with other independent bands when fleeing, only to drift apart soon after making good their escape. In the past, this trait was used to advantage by mustangers, who made large captures as their quarry herded together in flight from pursuing trucks and airplanes. Herding behavior served the horses better against natural enemies. When prairie wolves hunted the plains and sought foals as prey, a large number of stallions and mares were able to fend them off, forming a ring around their young and matching hoofs against the predators' fangs.

Only a trace of Spanish blood still flows in the veins of the crossbred descendants on the wild horse ranges of today, though a few still give hint of their pure-blooded ancestry. Markings such as faint stripes on the legs or withers are characteristic particularly of the horses in the Pryor Mountain refuge. Although it can be said that wild horses are no longer pure-blooded, it must also be added that there is endless beauty and surprise in their variety—a

13

PHOTOGRAPHIC LOCATIONS

NEVADA

1 Owhyee Desert
2 Trough Mountain
3 Sonoma Range
4 New Pass Range
5 Pine Nut Mountains
6 Stone Cabin Valley

WYOMING

7 White Mountain
8 Chain Lakes Flat
9 Atlantic Rim
10 Adobe Town district

MONTANA

11 Pryor Mountain
    Wild Horse Refuge

diversity of color that sets them apart from all native wildlife.

Bays comprise about half of the horse color in most areas, while in many districts of Nevada at least eight of ten are a reddish brown or dark tan. Blacks and sorrels follow in abundance. Greys and roans are fewer in number. Pintos are plentiful in Wyoming, especially on Chain Lakes Flat; they are uncommon in Nevada, however, except in a few remote areas. The Medicine Hat color phase, the "war bonnet" horse so prized by the Plains Indians, is very rare. Palominos and buckskins are scattered, but are more common in Nevada than elsewhere.

Appaloosas are also scattered. Both typical and leopard-spotted greys are found at Chain Lakes Flat in Wyoming, and Sonoma Range, Nevada. A few are also in the Aspen Mountains of Wyoming. Selective breeding by the Nez Perce Indians in the Palouse region of Idaho and Washington helped to maintain their distinctive color and variety.

Albinos are often present in the Sonoma and New Pass ranges of Nevada, while tarpan types (duns) live in Wyoming and southeastern Oregon.

Wild horses tend to be smaller in size than their domestic forebears. A thousand pounds is large for a wild stallion. This is a reflection of their remarkable adaptability to less than ideal environments. On the poor sagebrush desert ranges they inhabit today, these horses are unable to attain the stature of their stabled, hay- and oat-fed relatives. Stone Cabin Valley and its

environs in south-central Nevada provide an extreme example of this type of grazing land. Here, a population of some twelve hundred exists on one of the most dry and sparse of all the wild horse ranges. Overgrazing has compounded the natural harshness. Many horses are scrubby, their coats dull. Yearlings are stunted, and colts and fillies appear to be maturing at a slowed rate. Numerous adults are noticeably undersized. In general, I have found that the horses of Wyoming benefit from somewhat better habitats than those in Nevada, and are correspondingly larger and cleaner in appearance.

Smaller size, however, in no way indicates weakness or sickliness among wild horses. In fact, quite the opposite is true. Generations of survival on the American plains and desertlands have restored the natural durability and lean strength of their untamed ancestors. Theirs is the hardiness and stamina of the zebra ranging across the African savannah and the wild Przewalski's horse of the bleak Mongolian steppes. Few domestic horses are a match for the speed and endurance of a wild stallion in his prime. On all ranges, even the poorest, any dominant stallion who has proven his superiority over his rivals is always an impressive specimen.

Diminished stature has been proven to be an environmental adaptation, as offspring of horses captured in the wild generally develop to normal height and weight when raised under favorable conditions.

Living in their natural state, a horse's keen senses and wariness are immediately apparent. Equine sight, though not the equal of the binocular-like vision of the pronghorn antelope that often live on the same ranges, is nevertheless sharp in detecting movement at long distances.

Their colorblind eyes must see the world in a rather flat manner, poor of detail, for by sitting dead still, with my back to a boulder or bush, I've gone completely unnoticed as horses passed within a few yards of me.

Like domestic horses, they have a wide field

of vision, enabling them, with just slight movements of the head, to watch for danger approaching from any direction. They are able to focus both on the ground and out into the distance while grazing. Rival stallions often make stealthy approaches toward a straying mare; while appearing only to be feeding, their eyes never leave her or the band stallion. Their night sight is excellent, too.

The broad visual range of these wild horses is enhanced by an acute sense of hearing and smell. Ears twist and turn independently, recording the slightest sound. Earthbound vibrations, such as the distant approach of a galloping herd, are transmitted upward along the leg bones. Their quivering nostrils easily detect the scent of a human nearing from upwind, or a previous human presence. Once, I broke off the top of a tall sage bush blocking my view of a water hole only to have my carelessness ruin a chance for some good pictures, as when an hour later a band approached to water. Before drinking, the lead mare's sensitive muzzle located the scent from my palms mingled with the pungent aroma of the broken sage. Pausing only to turn and pass her droppings on the offensive odor—the only instance I've witnessed of a wild horse covering human scent—she sounded a warning and led the others in retreat.

That unusual incident illustrates the alarm invariably displayed by wild horses at human intrusion. How quickly fear prompts them to flee, however, varies from place to place. In some areas, a stationary car or truck may only arouse suspicion and draw skittish investigation from a safe distance, as would the sudden appearance of any foreign object. In remote areas, though, most horses begin to flee as soon as a moving vehicle appears over a hill as much as a mile away. Nearer to a ranch road, a band might tolerate the slow passing of a pickup but turn and run when it stops. A person on foot draws differing rates of reaction as well. On nearly inaccessible ranges, the horses may stare in overwhelming curiosity at the upright, two-legged creature before finally running away, but they will likely retreat at first glance the next day, and if followed, abandon the area entirely for a time. At the opposite extreme, several bands at Chain Lakes Flat regularly allowed me to slip within sixty to seventy-five yards before starting to edge away. Bands with foals are spookier than those without.

Wild horses are not often alarmed by nonhuman creatures. Coyotes are the largest predators remaining on most of their ranges, and those that I have seen skulking near bands—probably searching for small rodents disturbed by the horses' hoofs—have been ignored. In the Pryor Mountains, however, one mare with a tiny foal did react nervously to a coyote's howl, shifting about and staring into the distance. The call perhaps stirred instinctive fears of larger animals who hunted in wild horse country not so long ago.

Because of their fine night vision, horses are

often active after dark, and though they sleep somewhat more at night, the other routines of their lives continue as during the day. As a result of my limited nocturnal observations, I have also concluded that wild horses are bolder after nightfall. They often came near my camp under the cloak of darkness, and one moonlit night a young grey appaloosa stallion, who had shyly kept beyond camera range for days, drifted ghost-like to within twenty feet of me as I leaned motionless against my truck. Amorous wild stallions also approach ranch mares after the stars appear and have succeeded in spiriting many away.

Feeding occupies about half of a horse's time, as each consumes as much as twenty to twenty-five pounds of vegetation daily. A Wyoming study revealed that horses in the wild consume twenty-eight forage species, primarily grasses and grass-like plants. At times they also nibble on sagebrush, and I have watched them deftly pluck the tops from thistles and then tear the prickly plants from the ground at the roots with their hoofs to feed on the soft, juicy underpulp. Grazing requires considerable movement in their arid habitat, as they travel from one small clump of dry grass to the next. In times of extended drought, when the already meager food supply of the desert is reduced, they have been reported to supplement their diets by feeding on old droppings.

They seek out mineral licks regularly. In the Pryor Mountain refuge, the banks exposed by old mining activity are favored, and shallow cavities have been worn into the soft rock by hundreds of licking tongues. A drink is sometimes taken from the mineral-rich pools that usually collect just inside a horizontal shaft. Watering occurs twice a day on the average in the summertime, more often if the water is nearby, and usually once daily during the cooler months. If disturbed, wary bands will only drink at night.

The available water on a horse range can often be found by following converging horse trails. From high above, the trails spreading outward from a water hole sketch a faint tracery on the desert soil. A trip to the water hole offers a few moments of excitement for the horses, as well as a chance to stretch their legs. Often, after a foal or yearling darts into a galloping dash around the band, kicking up its heels and shaking its short mane with youthful energy, all the members will break into an exuberant trot as water comes into view. The final approach is always cautious, with the stallion or dominant mare leading, and the rest trailing slowly behind in single file. They pause often to scan for danger or the presence of other bands. Circling, all senses alert, the stallion may trot forward first while the others look on for a last assurance of safety. Or, if all are confident, the mares will bunch behind him and then surge past, arriving at the water's edge with a rush of drumming hoofs.

After a quick drink, some bands will leave

the vicinity of the water hole immediately, the stallion wary of meeting a rival at such a popular place. Others water leisurely, wading in up to their bellies as they drink, standing for a few tranquil moments in the shallows afterward. Where the water is deep enough they may swim briefly. Stallions in particular enjoy roiling the water into a clouded swirl with a forehoof and then dropping to roll and coat their hides in cool mud. The ground beside a water hole is also a favorite dusting place, and a large patch along the bank may be worn bare from repeated

19

use as a dry wallow. For young stallions, the open ground is perfect for a bout of play fighting.

Very dominant males may act to hoard a water hole, claiming it as their personal territory for as long as an hour at a time. While their mares graze or sleep beside the water's brink, they drive back any late arrivals.

Though bands utilizing the same water hole tend to stagger the times of their visits, sudden meetings, followed by friction between any stallions present, are inevitable. Many threat displays and serious battles occur beside the still surface of an isolated pool.

Over the past hundred years, stockmen have inadvertently opened large areas of once waterless, uninhabitable range to wild horses. Today, it is a safe guess that more horses drink from manmade cattle ponds than from natural sources on western rangelands, though they seem to prefer the latter when a choice is available. In those locations where cattle are not run, the horses must sometimes dig for water in dry washes when natural sources dry up in the late summer.

While mud baths and dusting contribute to their physical well-being by rubbing loose particles from the hide and fluffing matted hair, horses find greater satisfaction in mutual grooming with its added social benefits. Such grooming is usually done in pairs, involving either the closest companions in a young stallion band, a mare and her foal, two mares of similar rank within a band, or foals and yearlings. Dominant stallions rarely groom with another horse; when they do, it is usually with a favored mare. The pair stands side by side, facing opposite, nipping along the partner's neck, mane, withers, or back, and at the root of the tail; this may go on for up to several minutes. Then both halt at the same time and stand quietly together, itches relieved and friendships reinforced.

Horses enjoy several different levels of sleep intermittently during the day and night. Dozing in a relaxed standing position, head drooping and one hind leg flexed so that only the hoof tip is touching the ground, is the most common form of rest for adults. The members of a band are often clustered together in this attitude during the heat of the day or after a period of grazing. Upright and ready to flee approaching danger from the moment of awakening, it is a secure stance.

Lying down with legs tucked under the body provides deeper relaxation. This position is practiced frequently by mares, but few dominant stallions risk resting in this manner for long lest a challenger draw too near unobserved. Many mature males dare not experience such deep sleep; for years they only doze.

When lying stretched out on their side, horses are in their deepest state of sleep. A foal rests most often in this fashion. Awakening to find that its mother has grazed a short distance away, out of sight, it will quickly rise and whinny in a terrified voice until she answers,

then hurry to her side. Mares and young stallions also sleep in this manner when at complete ease.

Equine communication is accomplished partly by voice but more often through gestures, a great many of which are too subtle for the human eye to readily perceive. A wide variety of facial and body signals form a complete visual language between proximate horses, and often, many gestures are used in combination. A stallion threatening another will flatten his ears, arch his neck, and roll his eyes all at the same time. Ear positions are a signal of mood for all horses. Certain stances also give clear meaning; for example, bunched hindquarters and tensed hind legs warn of a kick. Head bobbing shows nervousness or agitation.

Voice is, of course, more apparent to human observers, especially when the shy nature of wild horses limits close study. Several sounds are heard more frequently in the wild than from the calm world of fenced pastures. The snort of alarm—a quick inhalation and sharp exhalation— at the discovery of an intruder is often repeated a number of times in rapid succession by mares and stallions alike before a band turns to flee. Then the stallion may halt, square broadside to look back, and repeat the alarm as his harem outdistances the danger.

The battle cry of enraged males is a sound that splits the air. Perhaps best described as a roaring squeal, it cuts off short at the end, as if the stallion's breath is abruptly exhausted. The long-drawn neigh of challenge between distant rivals is like a call from the depths of the primitive past.

Certain equine behaviors are, with rare exception, displayed only by stallions. Most are related to the possession of and dominance over a harem. Though individual males differ in the intensity with which they execute their dominant roles, a stallion is always the most active member of a band, for maintaining its integrity is the paramount task in his life. Some individuals appear to guard and herd their mares almost continually. The herding posture is distinctive: neck outstretched, ears pinned flat, head low and wagging snake-like from side to side, each dominant male directs his band and keeps the females and offspring bunched.

Marking is a regular duty. Particularly during the breeding season, a band stallion who sees one of his mares pass droppings or urinating is prone to urinate on the spot himself. His instinctive intent may be either to leave a message for other stallions that this female is one of those in his possession or to mask the scent of her coming into heat.

Stud piling is associated with the display of dominance among males. Mounds of feces several feet across and a half a foot or more deep are common beside water holes, where horse trails intersect, at the peak of saddles on well-traveled routes through hills, even on back country roads. Stallions are likely to pause at any stud pile they encounter, first sniffing to detect the scents of neighboring males, then adding their own sign, and then smelling a second time as if to judge the effect of their message. When two or more meet at a pile, elaborate posturing accompanies the ritual passing of droppings. Subdominant males mark first, dominants mark last. The order of ranking among the members of a bachelor band can be determined with certainty by their order of marking.

In flehmen, a stallion raises his head on outstretched neck and curls his upper lip high to

draw scents into a secondary olfactory center deep along his nasal passage. By this means, it is believed that the odor from a mare's urine is tested to determine whether or not she is approaching estrus. Male foals may flehmen in response to the smell of their dam or other mares when only days old. Unusual or potent odors will sometimes prompt a mare to flehmen, as one once did when she caught the first whiff of my scent at close range.

At times stallions show a curiosity toward unknown objects that can be unnerving to inexperienced observers. A crouched or still human may draw an aggressive investigation by a suspicious male intent on discovering any lurking foes. The horse's belligerence can be misinterpreted as a "charge." When he halts to posture and display—pawing the ground, shaking his mane, depositing ritual dung—the passing seconds can seem like hours. But all is bluff and show. Once the observer displays no threat response in return or is recognized as human, fear takes over and the stallion, riled to near fury a moment earlier, turns tail to flee, leaving a wake of dust.

It is a measure of a natural desire to avoid decimation of the breeding population that the majority of stallion disputes are settled through noncontact displays. Most dominant stallions purposely avoid young stallion bands in order to minimize the number of foes they must face. When actual fighting does occur, erupting in a split-second without preliminary ritual, it can be contagious, one battle prompting another among different foes in the vicinity.

While photographing at a Nevada water hole, I saw nothing more than threat displays between the local males for several weeks. Then one morning a feisty, lone newcomer appeared. An air of aggression suddenly overspread the valley. Within an hour, in addition to the two bloody contests this grey stallion became involved in, three other pitched skirmishes broke out among different nearby males.

Such contagions of aggressive behavior have the strongest influence within multiple-stallion bands. There the balance of rivalry is already so precarious that the mere sound of a distant battle is often enough to spur the subdominant male into another challenge. The sight of a nearby mating can have the same effect. Even such events as the approach of a thunderstorm, the passing of a low-flying plane, or the sudden presence of a car or human may spark them into fighting.

Most wounds suffered during a fight are superficial, but broken bones may result from full battle, sometimes leaving a stallion crippled for life. Rarely does death occur from these confrontations. Rearing presents the most vulnerable position, for balance during those high-reaching moments is at its most strained, and a fighter knocked backward can be severely injured.

Defeated stallions are quick to vent their frustration and disappointment. If a young stallion band is close by, chasing the bachelors is a favorite manner of restoring a damaged ego. After battle, many stallions like to roll in the dust to staunch their wounds.

It would be expected that by winning mares at random, all of the family bands within a single range would tend toward a color mix among their members appropriate to the general population. But this is often not the case, so often in fact that it is difficult to view the many exceptions simply as coincidences. In the opinion of most experienced observers, some stallions

show a definite preference for mares of a certain color or pattern. Thus, on colorful ranges, bands are sighted that are exclusively bay, or predominantly pinto where pintos are rare. And since wild stallions drive their daughters from the band, examples of color harmony are not inbred.

Most also have a favorite mare among the harem, the only form of friendship enjoyed by a dominant stallion. Unless otherwise occupied, the stallion remains nearest to her. In one large band, the white stallion's favored mate was the only

25

light-colored female in the harem. They stood out distinctly as they always grazed together. One theory suggests that favored mares or color preferences are determined by a stallion's memory of his mother's appearance.

Though stallions clearly exercise the dominant role in wild horse society, mares are not completely submissive. Individuals differ, and a few females display remarkable independence at times. An inexperienced stallion, just beginning to establish his harem, may acquire an elder mare with little patience for his enthusiastic herding and gestures of dominance. After feeling the sharp cut of her hind hoofs against his chest several times in rebuke, he is likely to find his life easier by tempering his behavior. Likewise, threats of or actual kicks and biting are used by many mares to discourage a sexually aggressive male when they are not yet receptive, or when being annoyed by any precocious colt within the band.

I once witnessed a large mare in the Pryor Mountains who, in the company of her foal and a lean filly, managed repeatedly to resist the intentions of a dominant stallion to herd her into his band. Clearly she was within his territory, and probably bore his offspring, yet her preference was to graze just within sight of his band at a distance of a quarter mile. Only when danger threatened did she join the band; on the following day she and her small entourage were separate again.

No matter how tyrannical the band stallion, a mare with a newborn foal is quick to warn him —and most other mares—from drawing too near, though wild stallions almost never harm their offspring. Instances of extreme anxiety involving her young may lead to outright rebellion.

Several days after I observed a lone stallion take a mare and sickly foal from a large Wyoming band, I watched as the foal slumped to the ground for the last time and died a short while later. The stallion, showing no concern for another male's dead offspring, soon attempted to drive the mare away. She refused to leave the motionless little form that lay sprawled at her hoofs. Like challenging males, the two confronted face-to-face, breathing loud threats into each other's nostrils. Then, with a scream of fury, the mare whirled and lashed out at the male's chin, and he backed away. Her rebellious attitude led to a standoff lasting a full afternoon, as she resolutely remained near her lost foal. Both were gone the following morning. It has been reported that mares whose foals have died become "colt crazy," willing to nurse other youngsters in the band and occasionally attempting to steal one from another member of the harem.

Lead mares sometimes take a partial role in the defense of a harem's well-being. When a family band and a young stallion band meet at a water hole, and the band stallion is preoccupied with the bachelor leader, the dominant mare will chase the other young males back from the

water so that she and the rest of the mares can drink in peace.

After an eleven month gestation period, most wild foals are born during the night. They are able to stand and nurse within the hour, and quickly grow steady enough on their legs to travel. One spring morning in northern Nevada I found a mare and her hours-old sorrel foal hiding in a small gully. Certain that the foal was still weak and I would be able to move near for some close-up pictures, I was startled when the mare led him away at a fast walk. They easily kept well ahead of me and rejoined their band without allowing me near enough to shoot a single photograph.

Foals are the liveliness of a family band. While the adults sedately graze or rest, a youngster will suddenly launch into a mad gallop around its mother and then race off through the band into the open sage beyond, running in its peculiar ham-strung manner, tossing its head and kicking up its heels in sheer delight. One may be struck by the urge to disturb another foal or older sibling, tugging at the mane of the resting brother or sister until the tormented sleeper is made to rise.

Friendships are most likely between foals of the same sex, fillies being less rambunctious than the colts. Mutual grooming is enjoyed between pairs. The play of young males—chasing, mane tugging, tail biting, half rearing—shows unmistakable signs of movements that will later be used in serious fashion. Yearling colts are often drawn into the fun, but play too rough and soon send the foals hurrying to the protection of their mothers.

Foals remain at the dam's flank during the mating season, when the youngsters are about a month of age. By watching, they seem to learn of courtship and mating. As they become older, some remain closer to their respective mothers, often rubbing against her and resting a tired head on her back, while others tend to stray farther off with friends. Wild foals are weaned by a year of age, but I have seen yearlings suckle on occasion without any reproof from the dam.

Young horses commonly display submission toward their elders through a curious gesture. After extending their neck so that their muzzle is near the adult's, they part their lips and work the lower jaw up and down in an action termed teeth clapping. I've observed this signal most often between colts and their fathers, either when the young male is still in the family band or after he has left and the two happen to meet.

Wild horses, mature and juvenile alike, must withstand winter snowstorms and violent summer thundershowers without the benefit of natural shelter. Though often restless at the storm's approach, in most instances they simply stand with rumps to the wind and quietly endure the fury until it passes. But at times a near flash of lightning and deafening crash of thunder sends them into a panic. From the safety of my truck I once watched an entire herd split apart in the midst of a fast-traveling curtain of lightning bolts. After stampeding ahead of the storm, they halted near my camp in a scene of frantic excitement. Individuals jumped and bucked in a wild swirl, their frightened cries sounding over the fury of the storm. A pair of stallions reared and dueled, lit by a brilliant flash, then raced away once again, in full flight under the falling rain and darkness. Where possible, horses will move from flat areas into nearby hills during storms of long duration. Perhaps it is during such terrifying moments that missteps are

28

taken and legs injured on rough ground or by stepping in the burrows of small animals. Limping horses are a regular sight wherever I have photographed. Fortunately, most do seem to heal and recover.

Domestic horses live longer than their untamed cousins, as is to be expected. The rigors of living in the wild shorten the life expectancy to about twenty years on the western deserts.

29

Nearly a century and a half has passed since wild horses reached their peak in America. They roamed through the tall prairie grasses by the millions, alongside bison numbering in the tens of millions. Little disturbed by Indians or the occasional passing of a white trapper, their domain encompassed countless miles within the length and breadth of the plains. Then came a long period of decline that brought them near extinction in North America for a second time.

Westward expansion by land-hungry settlers and ranchers encroached on the territories of Indian, buffalo, and wild horse alike. Relentless growth of the early cattle industry had perhaps the most dramatic effect on the wild horses. Overstocking of the range led to rapid deterioration of the grasslands, and wild horses provided a convenient scapegoat for cattle barons, so blinded by greed that they could not see their own mismanagement of the prairies. Accused of being responsible for spoiled public grazing land, contemptuously termed "feral," tens of thousands of horses were hunted down and shot. In the 1880s, ranchers in Wyoming paid a bounty for each wild stallion killed. Though not slaughtered as precipitously as the buffalo, those that survived the cattlemen's guns learned to range elsewhere and fell back to quieter, if less desirable, pastures.

The harsh winter of 1886–87 caused cattle losses in excess of eighty percent and bankrupted western stock interests. The tide of land use was only momentarily halted by the catastrophe,

however, and the horses continued to be forced onto poorer ranges; their numbers were further reduced during the following decade.

By the first years of decline, the dilution of the original Arabian-Barb bloodstock had already begun. Abandoned or escaped stock of cavalry, then of settlers and ranchers, promptly joined with their wild cousins and added their genes to the existing population. This addition of domestic horses continued to the twentieth century, reaching its height in the "dust bowl" days when many animals were turned out on public lands as water and forage became scarce.

A long-time practice of some ranchers also added to the mix through the years. Enterprising stockmen found an easy source of income by turning loose a well-bred stallion among neighboring wild horses, then capturing the crossbred offspring for sale as superior mounts.

The next great reduction in wild horse numbers involved the mass capture of live animals. Wartime demand for horses, especially during the Boer War (1899 to 1902) and World War I, gave rise to the practice of mustanging throughout the West. Horses were caught in any manner possible – driven into box canyons or hidden corrals, lured into traps surrounding water holes – and sold for use on foreign battlefields. Capturing wild horses could be dangerous work. Stallions snarled in the mustangers' ropes were known to attack with such ferocity that only a bullet would subdue them. By the 1920s perhaps only one million horses still ran wild.

With the passage of the Taylor Grazing Act in 1934, the government became involved in wild horse removal. The United States Grazing Service (later to become the Bureau of Land Management) was created to administer the vast tracts of public lands in western states. Heavily influenced by the interests of cattlemen, large government roundup programs depleted one hundred thousand horses from Nevada alone.

After World War II, a commercial demand for horsemeat as pet food made horse hunting once again profitable. Mustanging became mechanized and efficient. Airplanes were used to rout previously unmolested bands from their remote sanctuaries. Once on flat land, trucks coursed them to exhaustion. Then, often unfed and unwatered, they were shipped to distant processing factories.

By 1967, the Bureau of Land Management estimated that only 17,000 wild horses remained on public lands, their numbers declining sharply. Many feared that these living symbols of our past would soon vanish forever.

A partial step in the protection of the horses was taken in 1959. After many years of effort on the part of a dedicated few, led by the late Velma Johnston of Reno, Nevada, against the cruelty of the mustangers, the first federal law to recognize the existence of wild horses and burros was passed. Dubbed the "Wild Horse Annie Act" for the nickname the diminutive Mrs. Johnston adopted after overhearing a senator make reference to her at a Congressional hearing in Washington, D.C., it prohibited the use of motorized vehicles and aircraft in chasing, harassing, or capturing the animals on public lands. Enforcement of the law was difficult, and horses could still be captured and destroyed under state or local law. The decline continued.

For the next dozen years, Wild Horse Annie, aided by humane societies and horse protection organizations, continued her campaign to save the animals. True and complete protection for the horses was not achieved. Then, on the strength of a nationwide letter-writing crusade by thousands of school children, Congress finally responded a second time with passage of the Wild Free-Roaming Horse and Burro Act in 1971. The act prohibited unauthorized removal of the animals from federal lands, and delegated responsibility for them to the Department of Interior through the Bureau of Land Management (BLM).

Once granted the full protection of the law, unharried by man at last, wild horse populations began to rebound and then soar. Within a decade their numbers swelled to a thriving population of over 55,000.

Without natural enemies (only disease, severe weather, and high foal mortality limit their increase), wild horses now exhibit a growth rate of fifteen percent annually. This veritable population explosion, at times threatening major damage to the sparse ranges they inhabit, has led the Bureau of Land Management to institute a control program. Presently, what are determined to be excess numbers of wild horses are rounded up by helicopter, an economical method proven safe for horses and wranglers alike, and then adopted out to qualified applicants through the much publicized "Adopt-A-Horse" program. But the control effort, seen as necessary by many and justified by the BLM as a part of its effort to balance the needs of wildlife with human endeavors on public resource lands, has been marked by storms of controversy and litigation. As in the past, many ranchers who lease grazing rights on the sagelands view significant numbers of wild horses as a threat to their livelihood. Alternatively, some wildlife and horse protection groups oppose the roundup operations entirely or take exception to the BLM's decision as to how many horses should be allowed to remain as a representative number (on some ranges government studies called for a reduction of up

to eighty-five percent of the existing horses). Some people simply feel that, as one Wyoming resident told me, "If those horses can make a living out there on the desert, then they should just be left alone."

Regardless of the arguments that continue to swirl around them, wild horses appear to have a permanent place in our future. At the time of this writing, some 45,000 are roaming free over ten western states. More than three-quarters live in Nevada and Wyoming. Most of the remaining horses are found in southeastern Oregon, the extreme northeast of California, and Utah. Populations of less than one thousand are present (in order of most to least) in Idaho, Colorado, Montana, Arizona, and New Mexico.

Should you wish to view the horses for yourself, contact Bureau of Land Management offices in these states for specific directions and access information to local herds.

ROAMING FREE
Approximate Range
of Western American
Wild Horse Herds

OREGON

MONTANA

IDAHO

WYOMING

NEVADA

UTAH

COLORADO

CALIFORNIA

ARIZONA

NEW MEXICO

The early morning sun softens the weathered sandstone ranges that thrust upward from the desert floor. This is Wyoming badland country, eroded and shaped by wind and storms. The horses live in rolling valleys among sun-faded hills and cliffs. Their trails wind through the crumbling bleakness from one territory to the next, and they share what little pasture and water there are with no other large animals but antelope and occasional wandering deer.

35

On the Nevada wild horse ranges of Sonoma and Trough Mountain, most of the horses live up in the foothills of rocky ranges. Some follow retreating snows upward in the spring until they near the cliff crags, then descend once again in autumn. Others prefer the lowest slopes year-round, where bushy mountain mahogany trees offer shade.

37

38   The largest herds, however, roam wide valleys and basins and grey-green plains that run level to the horizon in the Owyhee Desert and Stone Cabin Valley. Immense oceans of sagebrush are broken only by the canyonlands of a few small rivers like the Little Humbolt, and dotted with the dry beds of Button and other seasonal lakes.

Wild horses are here, somewhere near, their hoofmarks imprinted in the thin dust. Their trails, worn fetlock deep, trace wandering patterns over the lonely land. Suddenly, they come into sight, distant forms in the heat-shimmer and barrenness. In moments, land and animals meld; this is their place, as untamed as the horses themselves, and less lonely and sad because of their presence.

41

Battling stallions, stampedes, lightning storms: these are moments of high excitement for wild horses. Many days and nights pass between such events when nothing disturbs the rhythm of the routines that make up a wild horse's life.

Feeding, watering, dusting, and sleeping take up their time; the business of survival is a serious one. They shuffle steadily over sparse pastures shared with pronghorn antelope and smaller native creatures, gleaning the thin grasses and dozing intermittently. The search for food takes up well over half their waking hours.

43

44 Water is their relief from thirst, heat, and insects. Wild horses at a water hole, drinking undisturbed, are a special joy to behold. There is always a hint of nervousness, though. Instinctive fears of predators cause quick plunging away at the slightest fright, real or imagined. Sometimes, when they can find them, horses water at streamlets which tumble from nearby rocky hills.

45

Dusting is often done immediately after watering, and mutual grooming combines physical comfort with social interaction. When they finally rest, few adult horses lay down—band stallions in particular cannot afford such relaxation. Shade is a rare treat on most wild horse ranges; when it is available, it is enthusiastically taken advantage of.

Young stallions, crossing the years from colthood to maturity, live a life of energy and nerves and budding instincts. The world is their playground, and movements of their lithe bodies in play or mock fighting are the very expression of equine youth.

With the grace of dancers and in a remarkable display of equine balance, the young pinto stallion and his bay companion embrace and neck-wrestle. Such playfulness is the nature of young males, a means of relieving excess energy, of honing physical skills, and acting out tactics of competition that will later be used in battle, as well as a manifestation of camaraderie within the bachelor band.

Most often, a round begins with a tentative mouthing of the partner's lips or a tug of the mane. The partner may answer this preliminary gesture by stepping away and declining the invitation, or by responding with a nip and tug of his own. The action promptly escalates, but the pace remains slow-motion in comparison to the lightning speed of serious combat.

Each bout is spontaneous. Regardless of the contestants, play is always marked by nimble but restrained action. Hoofs are held low while rearing, and kicks are only half-hearted and often pulled, directed so as to not make contact with hide. A few years later, as dominant stallions, there will be no time for frolic; the world is full of rivals. Horses have evolved behaviors for dealing with confrontation with a minimum of bloodshed. Threat displays maintain order, and fighting is a last resort.

Breathing heavily and glaring feverishly into one another's eyes, two stallions stand face to face. Seconds pass while they remain motionless—then the stillness breaks. With squeals of rage, in tandem, they rise to their hind legs, slashing the air. Forequarters hitting the ground, they separate and a truce is established.

Shows of strength do not always deter a rival and then fury displaces mere anger. Teeth and hoofs are brought into play and hide is gouged and torn. Stallions engaged in pitched battle are an awesome sight.

Rearing to bite or slash is only one of several forms of attack. Most of any battle is fought with all four hoofs on the ground, action lost from sight in a cloud of dust. Long fights that result in extensive but usually superficial injury are the exception. Heaving bodies, hoofs striking flesh, cries of fury: battling stallions are frightening to witness, particularly at close range. (I was once caught near a tangle of horseflesh and had a moment of panic before I recovered enough to move back, shooting photographs all the while.)

53

As would be expected, the most violent battles between stallions occur during breeding season. On one occasion, I observed a white stallion defend one of his mares and her filly twice in the same day. After dispatching his first competitor in a thundering confrontation, he was challenged again by a subdominant bay from within his own band. The white's wounds—received during the earlier battle—were barely staunched when he again took the offensive, landing a bite to the bay's unprotected neck. A running battle erupted across the sage, and the white was in danger of being dethroned for the second time that day. The bay proved unable to defeat the older stallion; however, unable or unwilling to drive him into permanent exile, the white allowed him to resume his second-rank position within the band.

55

In color and pattern the wild horse is unrivaled by any other creature of the sagebrush desert. They are the sound of passing hoofbeats in the night, a mare's soft nicker to her foal, the swish of tail and stamp of hoof in the afternoon heat. Nothing disturbs the wild horses and their solitude but whispers of desert winds.

Patience, stealth, and luck are needed to approach a wild foal. When very young, they are shy but too innocent to be afraid. With passing weeks, their natural wariness emerges and any strange movement or sound arouses suspicion. Intruders are confronted with a bright-eyed, penetrating stare.

59

I once found a foal who had somehow been separated from his dam. In desperate situations, a stallion will force a mare to abandon lagging offspring; or perhaps the mare had died. Timid and reluctant at first, the foal became bold when fed a mixture of powdered milk through the pinpricked finger of a glove. It was hard not to touch him (and thus give him the feared human scent)—he was so eager and innocent. Probably inadequate nourishment compared to mare's milk, the formula gave him enough energy to rekindle his latent instinct for survival. The last I saw of him, he was walking resolutely toward a distant grazing band.

Expectancy hangs in the warming spring air and the ground is tinged with the green of new growth. In every family band, mares are placid and heavy with their precious burdens. The foals arrive in a rush—spindly forms, eager for life. The mares hover near, a source of security in a very large world; foals look to their mothers for reassurance even into adolescence. The mares are also the source of nourishment and warmth: the demands of nursing, three or four times an hour in the beginning, are reluctantly met by some mothers, and a foal must sometimes rouse the mare for a meal. It is not long, though, before the young horses mimic the adults, and then demands for milk grow less frequent. Foals are quick to follow their mothers' every movement. In flight, a foal will run with its muzzle at her flank. I never tire of the sight of a mare and foal, for they radiate the beauty and peace of renewal.

It is always the stallion that is met first, face-to-face, for it is his duty to step away from the band and investigate any intruders. The stallion's life is one of fierce drive and hardship as well as magnificence. Once I came on the sad scene of a dying mare and her bewildered mate. He pawed gently at her bloated form, but she took no notice. Locoweed poisoning was the probable cause of her eventual death. The stallion remained with her body into the following day.

Stallions guard their bands with great zeal, and this sometimes leads to unusual and ridiculous incidents, such as a stallion chasing a bighorn sheep from the vicinity of his mares. As the lone ram, in a rarely seen inter-species relationship, continued to accompany this Pryor Mountain refuge band, the stallion's aggressiveness diminished. The horse herd seemed to offer partial satisfaction of the ram's social instincts, unfulfilled by his own species as the few wild sheep in this area are very scattered.

Sometimes an escapee from an area ranch will join the wild horse herd. After sorting out his proper place in their society, they will pry the halter off the convert's head, determined to remove the foreign symbol of the newcomer's past.

68   Young stallions are curious, flighty, mischievous characters. There are strong bonds between them, and they thrive on close contact. While staring at a crouching photographer, they crowd together—a signal of their need for the support of companions in any stressful situation. They are like a gang of small boys.

The equine family unit can be as small as three or as large as a harem made up of several mares and their offspring. Harems of matched coloration are not uncommon; such color harmony is too frequently sighted to be mere coincidence.

72   The largest herds of wild horses to be seen in modern times are in the Owyhee Desert of northern Nevada. Lush grass in early summer attracts large concentrations of the animals and the sight of them grazing in this remote environment is like a journey back in time. Wyoming also has impressive herds, particularly near Chain Lakes Flat in the south-central region of the state.

Unless frightened, horses are not usually seen moving at full gallop in the wild. Impressive, a herd's thundering stampede epitomizes the finest in earth-bound motion. Flashing hoofs, tousled manes, streaming tails—they pour past in a surge, a living, breathing, snorting flood of horses.

73

The air stills and cools. Hillsides are painted in the rich, soft hues cast by the falling sun. Valleys are dark pools of shadow. A tranquility settles over the land, while on the horizon the sky turns fiery.

Somewhere far across the sage a coyote sounds his sorrowful howl, and another answers. Overhead, a nighthawk glides past on fleet wings. This is the time when many creatures retreat into the safety and comfort of burrows while others are just emerging to prowl the night. Wild horses' lives go on, unchanged, in the descending darkness.

Lone horses sometimes run for the sheer joy of stretching their muscles. Traveling unhindered by fences, they range lonely miles, questing for a mate. May the prints of unshod hoofs always mark the land.

A hunter armed only with camera and film, Skylar Hansen learned the techniques of wildlife photography from Robert Kothenbeutel. This reinforced the lessons learned while growing up in Arizona and exploring the state's desert and mountains. Wild horses attracted his attention in 1971 and he has spent several years researching, making field observations, and photographing these magnificent animals.

Hansen's work can also be seen in Gibson's 1984 *Skylar Hansen's Wild Horse Calendar* and Hallmark's *Forever Free* 1984 calendar issue.